EGG

Photographed by
Jane Burton and Kim Taylor

Written by
Robert Burton

Dorling Kindersley
London • New York • Stuttgart

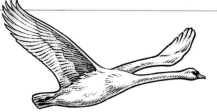

A DORLING KINDERSLEY BOOK

Editor Gillian Cooling
Designer Tina Robinson
Project editor Mary Ling
Project art editor Helen Senior
Assistant designer Susan St. Louis
U.S. editor B. Alison Weir
Production Samantha Larmour
Managing editor Sophie Mitchell
Managing art editor Miranda Kennedy

Illustrations David Hopkins
Ian Fleming and Associates Ltd.

First American Edition, 1994
4 6 8 10 9 7 5 3
Published in the United States by
Dorling Kindersley, Inc., 95 Madison Avenue
New York, New York 10016
Copyright © 1994 Dorling Kindersley Ltd., London

Library of Congress Cataloging-in-Publication Data
Burton, Jane.
Egg / photographed by Jane Burton and Kim Taylor;
written by Robert Burton. — 1st American ed.
p. cm.
Includes index.
ISBN 1-56458-460-7
1. Embryology – Juvenile literature. 2. Eggs – Juvenile literature.
[1. Embryology. 2. Eggs.] I. Taylor, Kim, ill. II. Burton,
Robert, 1942-   . III. Title.
QL956.5.B87   1994
591.3'3–dc20   93-28365
CIP
AC

Color reproduction by Koford, Singapore
Printed and bound in Italy by L.E.G.O.

CONTENTS

WHAT IS AN EGG?

AN EGG is the first stage in the development of a baby animal. It is made by a female animal. All eggs develop in the same way. They start as a tiny speck that grows into the different parts of the animal's body. When people think of an egg, they usually think of a chicken's egg – but there are many other kinds.

Falcon egg

Emu egg

Honeyeater egg

Bantam egg

COLOR
Many birds' eggs have brightly colored shells, which can be plain or speckled. Some shells have dull colors and patterns that make them difficult for enemies to see. This is called camouflage.

Thrush egg

This quail's nest shows how hard it can be to see well-camouflaged eggs in a nest.

SIZE
Eggs range in size from tiny eggs, such as insects' eggs, to very large eggs, like that of an ostrich. The dot below shows how small some insects' eggs are; some other animals' eggs are even smaller than this!

This is the actual size of a butterfly egg.

This is the actual size of an ostrich egg. It is 100 times longer than a butterfly egg, and is the largest egg in the world.

Leopard gecko egg

Leopard tortoise egg

SHAPE
Birds' eggs are always oval. Other animals have eggs that are different shapes, such as spherical tortoise eggs or tubular dogfish eggcases.

Dogfish eggcase

Partridge nest

NUMBER OF EGGS
Birds only lay a few big eggs. Other animals may lay thousands of small eggs.

Corn snake egg

EGGSHELLS
Shell thickness varies greatly between different animals' eggs. Birds and insects lay eggs with hard shells. A mineral called calcium makes birds' shells hard. Snakes and some other animals lay eggs with soft shells; frogs' eggs do not have a shell, but are protected by jelly instead.

Moorhen egg

Frog spawn

WHO HAS EGGS?

EVERY KIND OF ANIMAL makes eggs. Most animals lay their eggs. The baby animals develop inside the eggs and hatch out when developed. Other animals, including most mammals, do not lay their eggs. Instead, their eggs develop inside the mother's body, and are eventually born as babies.

BIRDS

Birds' eggs have a hard, chalky shell. Inside the egg, there is a thick layer of "egg white," or albumen, around the yellow yolk. This protects the egg from knocks. The baby bird grows on top of the yolk, which is its food until it hatches. Birds are the only animals that incubate (sit on) their eggs to keep them warm.

INSECTS

Most insects' eggs hatch as larvae and later become adults. Others hatch as nymphs and grow into winged adults.

MOLLUSKS

Slugs and snails lay their shiny round eggs in holes in damp ground. The yolk is surrounded by a layer of albumen and a hard shell.

AMPHIBIANS

Most amphibians' eggs develop in fresh water. The eggs are often covered in jelly, which protects them and keeps them warm.

REPTILES

Reptiles lay their eggs on land. The eggs have waterproof shells so they do not dry up. They hatch into miniature adults.

FISH

Fishes' eggs are laid in water and have soft shells. Some fishes' eggs hatch into larvae. A few kinds of fish give birth to babies.

THE DEVELOPING EGG

EVERY EGG starts as a single cell in its mother's body. It is fertilized by a cell called a sperm from the father. Once the egg is fertilized it starts to develop. The first stages of development are the same in most animals. The egg cell divides into two new cells, which divide again. Cell division continues and the developing animal, called an embryo, forms and starts to look more like its parents.

CHICKEN OR EGG ?

Which comes first? To get an egg you need a chicken, but to get a chicken, you need an egg! Female chickens, called hens, start laying eggs when they are about a year old.

FERTILIZING THE EGGS

The male chicken is called a rooster. When he mates with the hen, the eggs inside the hen are fertilized by sperm from the male. This must happen before a chick can start to develop inside an egg.

PREPARING THE NEST

Before the hen starts to lay her eggs, she makes a scrape in the ground with her beak and feet. She pulls twigs, feathers, hay, and leaves up around her. If one of the eggs rolls away she pulls it back under her with her beak.

THE EGGS

The hen lays between 7 to 15 eggs in a clutch. She lays one egg a day. She does not start to incubate them until the last egg of the clutch is laid. Because of this, all the eggs start to develop and hatch at the same time.

DEVELOPMENT OF A CHICKEN'S EGG

It takes 24 to 25 hours for an egg to form. This happens inside the hen, in a special tube called an oviduct. A newly laid egg does not have a chick inside it. The chick grows in the egg while its mother incubates it. She sits on the egg for 21 days to keep it warm, until the chick is ready to hatch out. If the egg gets cold, the chick may die.

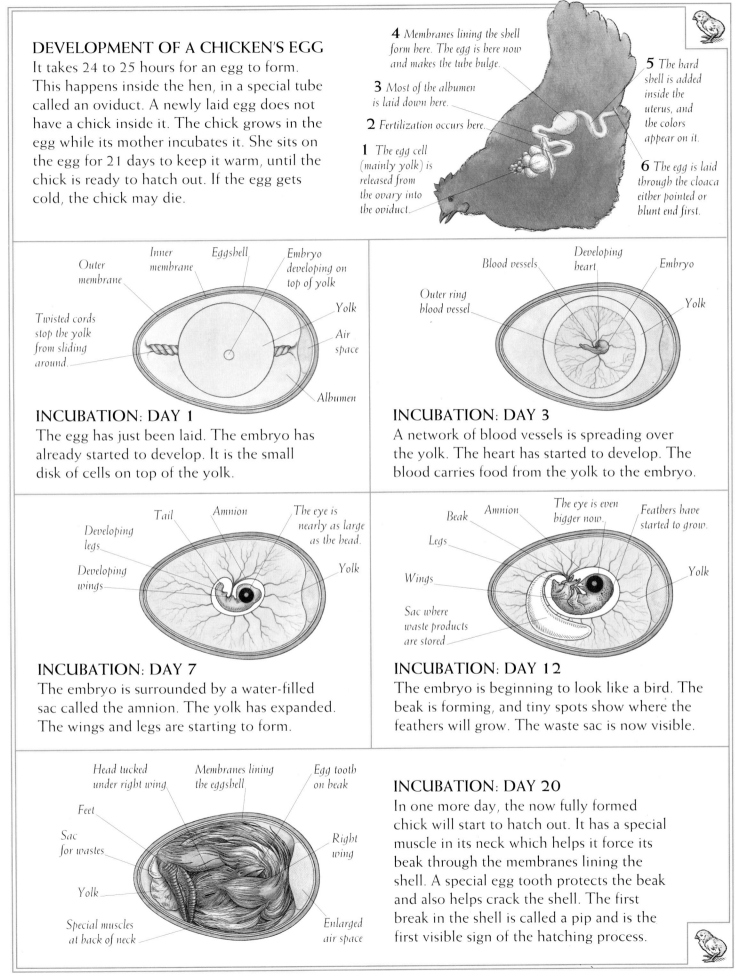

4 Membranes lining the shell form here. The egg is here now and makes the tube bulge.

3 Most of the albumen is laid down here.

2 Fertilization occurs here.

1 The egg cell (mainly yolk) is released from the ovary into the oviduct.

5 The hard shell is added inside the uterus, and the colors appear on it.

6 The egg is laid through the cloaca either pointed or blunt end first.

Outer membrane — Inner membrane — Eggshell — Embryo developing on top of yolk

Twisted cords stop the yolk from sliding around.

Yolk

Air space

Albumen

INCUBATION: DAY 1

The egg has just been laid. The embryo has already started to develop. It is the small disk of cells on top of the yolk.

Blood vessels — Developing heart — Embryo

Outer ring blood vessel

Yolk

INCUBATION: DAY 3

A network of blood vessels is spreading over the yolk. The heart has started to develop. The blood carries food from the yolk to the embryo.

Developing legs — Tail — Amnion — The eye is nearly as large as the head.

Developing wings

Yolk

INCUBATION: DAY 7

The embryo is surrounded by a water-filled sac called the amnion. The yolk has expanded. The wings and legs are starting to form.

Beak — Amnion — The eye is even bigger now. — Feathers have started to grow.

Legs

Wings

Sac where waste products are stored

Yolk

INCUBATION: DAY 12

The embryo is beginning to look like a bird. The beak is forming, and tiny spots show where the feathers will grow. The waste sac is now visible.

Head tucked under right wing — Membranes lining the eggshell — Egg tooth on beak

Feet

Sac for wastes

Right wing

Yolk

Special muscles at back of neck

Enlarged air space

INCUBATION: DAY 20

In one more day, the now fully formed chick will start to hatch out. It has a special muscle in its neck which helps it force its beak through the membranes lining the shell. A special egg tooth protects the beak and also helps crack the shell. The first break in the shell is called a pip and is the first visible sign of the hatching process.

OSTRICH

OSTRICHES LIVE in the dry savannahs of Africa. They are the largest birds in the world, and also lay the largest eggs. The female lays up to 12 eggs which can weigh 3½ pounds each. Ostriches are unusual because several females may mate with the same male and then lay their eggs in one communal nest.

OSTRICH eggs are 6 in. in length on average

THIS EGG IS ACTUAL SIZE

DAY 1 6.30AM
Ostrich chicks are normally ready to hatch after 42 days. When the baby ostrich is ready to come out, it usually turns around so that its head is positioned at the blunt end of the egg.

The shell is full of tiny holes. These allow air into the egg, so the chick can breathe.

DAY 3 8.10AM
With jerking movements, the baby ostrich uses its strong beak to hammer a hole in the shell. Because its neck is curled around, the hole is made near the middle of the egg. The shell of an ostrich egg is two millimeters thick and very strong. It can support the weight of two adult people standing on it! The parents may sometimes help the chicks hatch by pecking their eggs.

The chick begins to breathe properly as soon as the shell is pipped.

ONLY the most senior of these females sits on the eggs. There may be up to 30 eggs in one nest. The female shares incubation with the male. She sits on the nest during the day, and he sits on it at night. After they hatch, the baby ostriches gather in a group. The group is guarded by just one pair of adults until the young birds are a year old.

As the chick kicks, large pieces of shell fall off.

With powerful blows, the chick breaks the shell apart.

The eggshell is very thick.

DAY 3 9.30AM
Unlike most chicks, the ostrich chick has no egg tooth to help it get out of the egg. Instead, its beak and sturdy legs press hard to push a chunk of shell out and make the first hole. Then the chick kicks its way out with its powerful feet.

The ostrich chick continues hatching on the next page.

The baby ostrich has almost hatched out!

DAY 3 9.55AM
The ostrich chick has been hatching for more than 50 hours now. The hole it has made is almost large enough for it to squeeze through.

The chick's foot is sticking out.

Lining of eggshell

DAY 3 10.00AM
Suddenly the chick bursts out of the egg. After six weeks inside the shell that has kept it safe, the little ostrich tumbles out into the world.

Hatching has taken more than two days of hard work, and the chick is exhausted. Its downy feathers are still wet with fluid from the egg, so the adult birds keep it warm until it is dry. A few hours after hatching, the chick's feathers dry out and become fluffy.

The ostrich's ear is on the side of its head.

STANDING AT LAST

It is several days before the chick can stand securely. Then it will leave the nest and run around with the other chicks. Chicks from several nests join together to form a group called a crèche. The crèche may contain up to 100 chicks, but it is guarded by only one set of parents. The chicks find their own food, such as grasses, leaves, flowers, and seeds.

Ostrich feathers are fluffy, like hair.

The chick is three days old. It is bright and alert now. Ostrich chicks are fully grown after one year.

The chick lies exhausted after finally breaking free.

An ostrich has only two toes on each foot.

ROMAN GOOSE

THE ROMAN GOOSE is a small domestic goose. Most domestic geese are descended from the wild greylag goose. A Roman goose lays six creamy white eggs in each clutch, and up to 45 eggs in a year. These are incubated by the female for 28 days while the male, called the gander, keeps guard over the nest.

ROMAN GOOSE eggs are 3⅓ in. long on average

Hole in shell

DAY 1 4.00PM
The goose egg begins to hatch. A hole appears in the eggshell.

DAY 2 9.00AM
The baby goose, or gosling, starts to push the cap off the shell.

Gosling's head curled around

DAY 2 9.10AM
The gosling will soon be able to straighten its neck.

After more than 17 hours, the gosling is finally out. It is resting, curled up.

DAY 2 9.15AM
After resting for a while, the gosling will sit up.

STANDING STRONG
The gosling has large, strong legs when it hatches. In two or three days it leaves the nest and runs after its parents. They will lead it to a place where it can find food by itself.

The gosling is now one day old. Its feathers are yellow and fluffy.

MOORHEN

THE MOORHEN lives in many parts of the world. It builds its nest from a pile of plants at the edge of a river or pond. The nest is built high enough to keep the eggs out of the water – sometimes it even floats on the water. The female moorhen lays two clutches of five to nine brown speckled eggs.

> The average length of a MOORHEN egg is 2 in.

The eggs are incubated for 20 days.

DAY 2 12.00PM
A ring of holes has been made, a day after the egg is pipped.

Wing

Beak

DAY 2 12.30PM
By pushing hard, the chick opens up a slit in the shell.

DAY 2 12.32PM
The chick has almost pushed its way out of the egg.

The chick rests before it finally breaks free.

DAY 2 12.34PM
After more than a day's effort, the chick bursts out of the egg.

Wing with claw

FINALLY FREE
Moorhen chicks are soon strong enough to scramble and swim after their parents. When the weather is cold and wet, the chicks shelter under their parents. The chicks from the first clutch of eggs often help their parents look after the chicks from the second clutch.

The chick is now one day old.

This chick has no webbing between its toes. This causes it to swim jerkily.

BLACK SWAN

THE BLACK SWAN lives in Australia and New Zealand. It also lives in parks in other countries. Its nest is made from a heap of water plants, and can be up to a yard in diameter. Some nests float in the water. The male and female swans gather the plants and pile them up. The female then lines the nest with down to keep the eggs warm.

> BLACK SWAN eggs are 4 in. long on average

The eggs are normally pale green or brown.

Egg lining is exposed.

DAY 1 3.15AM
The male and female black swans sit on their eggs for about 35 days before the eggs start to hatch.

DAY 1 4.30PM
The cygnet makes a crack around the top of the egg with its beak. Small pieces of shell fall off.

The cygnet looks around now that it is out of its safe shell.

DAY 1 6.30PM
The cygnet is resting after spending most of the day struggling out of its shell. Its mother will keep it warm until its down becomes dry and fluffy.

THE FEMALE black swan lays five or six eggs. The parents take turns sitting on the eggs, which hatch in five to six weeks. The young are called cygnets. The oldest cygnets leave the nest with one parent before the other eggs have hatched. Once the cygnets have left, they never return. Sometimes several families of cygnets live on the same pond.

DAY 1 6.00PM
The crack now runs around the shell. The cygnet can start to push itself out.

Wing

DAY 1 6.15PM
The cygnet uses its legs to push itself out of the shell. Its wing is just sticking out.

FREE AT LAST
The cygnets' gray feathers turn black by the end of their first year.

These cygnets are now two days old.

MUSCOVY DUCK

MUSCOVY DUCKS come from South America, where they were tamed before being taken to other countries. Wild muscovies live in ponds and streams in forests, and feed on water plants and animals. They are good at perching on branches and nest in holes in tree trunks. The female lays 8 to 9 eggs that hatch in five weeks.

MUSCOVY DUCK eggs are about 2 1/2 in. long

DAY 2 9.00AM
The shell is in two pieces, 24 hours after the egg was pipped.

DAY 2 9.10AM
The duckling is pushing the top off the blunt end of the egg.

DAY 2 9.12AM
With a few more pushes, the duckling is almost out.

DAY 2 9.20AM
The duckling cheeps as it hatches out of the egg.

The duckling is strong enough to hold its head up.

The duckling is now a day old.

OUT OF THE EGG
The duckling yawns. This is not because it is tired or sleepy – it takes deep breaths to get its body working properly.

GOLDEN PHEASANT

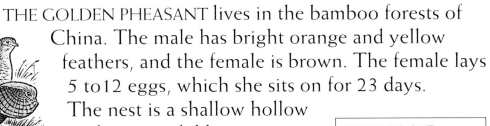

THE GOLDEN PHEASANT lives in the bamboo forests of China. The male has bright orange and yellow feathers, and the female is brown. The female lays 5 to 12 eggs, which she sits on for 23 days. The nest is a shallow hollow in the ground, like a saucer.

GOLDEN PHEASANT eggs are 2 in. in length on average

Beak with egg tooth

DAY 1 11.30PM
The golden pheasant chick makes the first hole in the shell.

DAY 2 4.30AM
It makes a ring of holes which join to form a long split.

Foot

DAY 2 4.40AM
The little chick uses its feet to help push itself out of the egg.

The chick's eyes are shut.

These three one-day-old chicks hatched out together.

DAY 2 4.41AM
After more than five hours, the chick is almost free.

FULLY HATCHED
Golden pheasant chicks hatch out of the eggs together. These three chicks are not yet ready to leave the nest. They will creep under their mother's body to keep warm.

JAPANESE QUAIL

THE JAPANESE QUAIL is a relative of the pheasant. The male's call can be easily heard, but the female, nest, and eggs remain hidden. The female lays 10 eggs. The nest is a shallow saucer, scraped in the earth, often among crops. The female's brown feathers camouflage her when sitting on the nest, keeping her safe from enemies.

The average length of a QUAIL egg is 1¼ in.

DAY 1 9.20PM
The quail egg starts to hatch 16 days after being laid.

DAY 1 10.37PM
The chick makes a neat crack around the blunt end of the egg.

DAY 1 10.50PM
The end of the egg opens like a lid as the chick pushes itself out.

The chick rests after its efforts.

DAY 1 10.55PM
It has taken 95 minutes since cracking the egg to hatch out.

UP AND ABOUT

Quail eggs in the same nest all hatch at the same time. The chicks cheep to each other from inside the eggs, and all start hatching together.

This chick is two days old. In real life, it is not much bigger than a bumblebee!

AYLESBURY DUCK

THE AYLESBURY DUCK is a white farmyard duck related to the wild mallard. It is named after the town of Aylesbury in England, which is famous for its ducks. The female lays 10 eggs in each clutch, and up to 100 eggs in a year. The female builds the nest, then lines it with down. The male does not help rear his young.

Most AYLESBURY DUCK eggs are about 2½ in. long

The eggs are white or pale green.

Egg has pipped.

DAY 1 5.50PM
After 28 days, the duckling is about to hatch out of its egg.

Beak

DAY 2 10.20AM
The young duckling makes holes in the shell with its beak.

DAY 2 1.55PM
The bird's neck is tightly folded to fit inside the egg.

DAY 2 1.56PM
With big kicks, the duckling struggles out of the shell.

After more than 20 hours, the chick is almost free. It rests while its down dries.

GROWING UP FAST
Two days after hatching, the ducklings leave the nest forever. The mother duck leads her hungry brood to the water to look for food.

The duckling is now two days old. Oil from its mother's feathers, and later from its own oil gland, makes its down waterproof.

HUMBOLDT PENGUIN

THE HUMBOLDT penguin lives in the sea off the coasts of Peru and Chile, and nests on the islands nearby. Penguins cannot fly in the air – instead, they "fly" through the water. Their sleek bodies are well adapted for diving and swimming, and their wings have evolved into flippers. They dive to catch fish, such as anchovies, to eat.

HUMBOLDT PENGUIN eggs are 3 in. long on average

The baby penguin has made three separate holes in the shell.

DAY 2 10.00AM
It is already 20 hours since the baby penguin made the first hole. It rests and turns inside the egg and keeps on making holes until a ring forms. All baby birds turn in the same direction in their shells.

The shell is now broken almost all the way around.

DAY 4 12.20PM
The chick has still not made enough holes in the shell, so it turns around again inside the egg and makes more holes. Eventually, it makes a jagged cut right around the blunt end of the egg.

After more than three days of effort, the chick has finally hatched out of the egg.

DAY 4 4.42PM
Suddenly the chick's neck uncoils, and its head shoots out of the shell. The lid of the shell flips open and the front of the chick's body comes out. Seconds later its sturdy feet kick backward, pushing the chick forward and kicking away the remains of the shell.

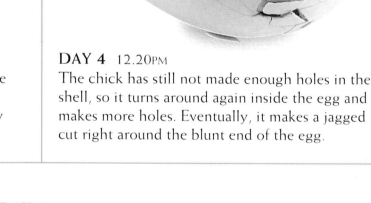

THE HUMBOLDT penguin comes ashore twice a year to lay its eggs. It lays two eggs in each clutch, which it incubates for 28 days. Penguins do not like hot weather, so they lay their eggs beneath rocky ledges or in burrows in the ground where the parents are shaded from the sun.

The baby bird can be seen through the widening crack.

DAY 4 2.40PM
Even though the chick is curled up in the shell, there is room enough for it to heave its wings and kick its feet. It also tries to straighten its neck. As it does, the crack opens up.

The penguin has a thick layer of fat under its skin to keep it warm. It also has special waterproof feathers that help keep it warm and dry, in and out of the water.

The chick is now three days old. Its down is velvety.

DAY 7
The chick is fed regular meals of fish by its parents. Seven weeks after hatching it will have grown a full coat of waterproof feathers. When it is 10 weeks old, the young penguin leaves its parents and swims out to sea.

WADDLING AROUND
The penguin is now two years old. The adults usually return to the place where they hatched to lay their own clutch of eggs.

BULLFINCH

THE BULLFINCH lives in woods and hedges. The adults eat mainly seeds and buds, but they feed insects to their young to provide a richer diet. The nest is a flimsy platform of twigs and roots hidden among the leaves of a tree or bush. The female lays as many as six eggs in each clutch, two to three times a year.

Most BULLFINCH eggs are about ¾ in. long

The eggs are normally pale blue with brownish purple speckles.

DAY 1 2.00AM
The egg is incubated for 12 to 14 days by the female bullfinch.

The pink body of the bird is just visible.

DAY 1 4.30PM
The chick has almost completed the ring of holes around the shell.

DAY 1 4.35PM
The nestling pushes hard to get out of its tight-fitting eggshell.

The baby chick is almost bald.

DAY 1 4.36PM
With a few last pushes the baby bird is almost free.

The nestling still has gray fluffy down on its back and neck.

DAY 6
The baby bird's feathers have started to sprout.

The bullfinch is a shy bird, but its bright colors make it easy to spot when it does appear.

The bird is now nine months old and is fully grown.

DAY 28
The young bird is not as colorful as its parents.

BRIGHT AND ALERT
When nesting, both the male and female adult bullfinches grow special pouches in their throats for carrying food to their young nestlings.

STARLING

STARLINGS ARE COMMON birds in town and country. They build their nests in holes in trees and buildings. The nest is a mass of twigs, leaves, and grass. The female lays a clutch of four to six eggs, usually once a year. She lays one egg a day, often in the morning. The female does most of the incubating, but the male sits on the eggs for a few hours a day.

STARLING eggs are 1¼ in. in length on average

The eggs are glossy.

DAY 1 6.20PM
Starlings' eggs range in color from white to pale blue.

Holes at blunt end of egg

DAY 2 9.15AM
The egg hatches after it has been incubated for 12 days.

DAY 2 9.20AM
As the chick starts to hatch out, its pink body can be seen.

Down

DAY 2 9.21AM
Its body is almost naked but it has some down.

In summer, the beak changes color from brown to yellow.

The starling is fully grown at nine months old.

During autumn and winter, the feathers have pale tips, which make the bird look spotted.

FULLY GROWN
Starlings are very adaptable birds and live in many parts of the world. They will eat most things they can find.

DAY 22
Young starlings leave the nest at three weeks old.

MUTE SWAN

MUTE SWANS build a huge nest of plants, often on an island or on the bank of a river or pond. The nest measures up to six feet across and 31 inches high. The male picks plants, then the female builds the nest and lines it with soft down. The female lays four to eight eggs and sits on them for five weeks.

> The average length of a MUTE SWAN egg is 4½ in.

Swan eggs vary in color from pale brown to gray-green.

The shell is pipped.

DAY 1 4.00PM
The first sign that the egg is going to hatch is a small break in the shell.

Egg tooth is just visible.

DAY 2 8.10AM
The baby swan, called a cygnet, cuts off the top of the shell by hammering it with its beak.

Egg lining soon begins to dry.

DAY 2 8.39AM
The cygnet often rests while it is hatching. The white egg tooth can be seen on the tip of the beak. It was used to cut through the eggshell, and will fall off after a few days.

After almost 17 hours, the cygnet is nearly free.

White egg tooth

Long neck unfolded at last

THE COB, or male swan, guards the nest until the eggs hatch. He sits on them when the pen, or female swan, goes to feed. The pen may hide her eggs by covering them with plants. Both male and female can be very aggressive, and may attack other pairs of swans or even humans who come too close to the nest. They fight with their strong beaks and wings.

Egg lining

Fluffy down soon dries.

DAY 2 8.35AM
The cygnet gradually straightens its long neck and pushes its way out of the broken eggshell.

The cygnet is now two days old. Its gray-brown down will turn white by the end of its first year.

GROWING STRONG
The young cygnet is ready to go swimming with its parents.

CROWNED LAPWING

THE CROWNED LAPWING lives on the grassy plains of eastern and southern Africa. It lays its eggs on the ground, and surrounds them with pebbles and dry grass. The eggs are safe because they are hard to see. The parents also scare away animals that come too close by screaming at them.

CROWNED LAPWING eggs are 1¹/₂ in. long on average

Crowned lapwings lay two or three pale eggs with dark markings.

DAY 1 10.00AM
The egg is almost ready to hatch after 25 days incubation.

The chick's beak can just be seen.

DAY 2 10.42PM
Hatching has started. The first pieces of shell have broken off.

DAY 2 11.09PM
The chick has burst open the shell and is pushing itself out.

The top of the eggshell is still on the chick's head.

The chick is resting but will soon be able to sit up.

DAY 2 11.10PM
The chick has straightened its neck and has thrust off the end of the eggshell.

The chick is now two days old. Baby crowned lapwings have very long legs and a coat of down.

LONG-LEGGED CHICK
Baby crowned lapwings soon leave the nest and run around with their parents, who show them where to find food. They can catch insects and other small animals by themselves.

STREET PIGEON

STREET PIGEONS live in large flocks in towns and cities. They are often joined by racing pigeons that have escaped and turned wild. Their nest of twigs is built on a ledge or in a hole in a building. Some pigeons eat food that people feed them and can become very tame this way.

The average length of a PIGEON egg is 1³/₄ in.

Pigeons lay two eggs in each clutch, which are incubated for 17 to 18 days.

DAY 2 10.05AM
The egg has now cracked, 17 hours after it was first pipped.

The baby pigeon can be seen through the cracks.

DAY 2 2.20PM
The cracks show that the chick is just about to come out.

Beak

DAY 2 2.35PM
The end of the shell is being pushed off by the chick inside.

White egg tooth

The squab will be fed for the first 10 days on a liquid made in the parents' throats called "pigeons' milk."

DAY 2 4.30PM
The baby pigeon, called a squab, is finally out of the egg. It is already dry and fluffy.

The young pigeon still has some yellow fluffy down.

YOUNG SQUEAKER
The young pigeon is now two weeks old. It is called a squeaker because it squeaks when it wants to be fed. It will be able to fly when it is 35 days old.

TAWNY OWL

MOST TAWNY OWLS live in woods, but some live in towns or open country with few trees. When they mate, they do so for life. The male and female live with each other all year round. The female does not build her own nest. Instead, she finds a hole in a tree or uses the abandoned nest of another bird.

TAWNY OWL eggs are 1¼ in. in length on average

The eggs are laid at the end of winter.

DAY 2 8.20AM
This egg was chipped more than 24 hours ago, after four weeks' incubation. The chick waits in the egg until the remains of the yolk are absorbed into its body. It breathes through the cracks.

The chick may call to its mother from inside the shell.

DAY 3 7.30AM
The chick presses its beak in the same spot until small bits of shell fall off, and a hole is made. Inside, the chick moves around to get into the right position for the final stage of hatching.

The chick's wet down shows through the slit.

DAY 3 12.20PM
The chick works fast to chip out a series of holes. The holes are so close together that they make a slit. The slit widens, and the end of the egg lifts away as the chick heaves with its neck and wings.

The chick is finally free, more than 55 hours after it first chipped its shell.

DAY 3 12.44PM
The mother owl will continue to brood the newly hatched owlet. The warmth of her body will soon dry its down. The owlet does not feed for the first day because it still has some yolk inside its body.

DAY 5
The owlet is now two days old. It keeps its eyes shut for most of the time until it is about two weeks old. It twitters when it is hungry and its mother feeds it with tiny shreds of mouse meat.

THE FEMALE lays two or three eggs, or more if she catches plenty of food. She starts to incubate as soon as the first egg is laid. During this time, her mate brings her food. The eggs hatch over a number of days, in the order they were laid. After all the eggs have hatched, both parents feed the nestlings.

The egg tooth can be seen clearly.

DAY 3 12.34PM
As the chick heaves, the slit widens until a wing escapes from the shell. The chick's head is still curled around. At each heave, the chick tries to straighten its neck and get its head out from under its other wing.

Owls' eyes face forward and cannot swivel in their sockets. Because of this, owls have to turn their heads in almost a complete circle to see all around them.

The tawny owl only comes out at night, but its well-known hooting often reveals where it is.

The chick still has its egg tooth.

WIDE-EYED
The 12-week-old owlet can fly and is learning to hunt for itself. But it still relies on its parents to bring it food. It can swallow a mouse - whole! When it is a year old, it will breed and have its own babies.

LEOPARD TORTOISE

ALL TORTOISES live in warm countries. The leopard tortoise lives in Africa. It eats grass and juicy plants. Most adults have shells that measure about 14 inches long, but one leopard tortoise was measured at 25½ inches long and weighed 95 pounds!

LEOPARD TORTOISE eggs are round and are 1¾ in. across

The baby tortoise is peeping out of the hole in the shell.

Lining of eggshell

DAY 1 8.30PM
The baby tortoise has started to hatch out. Its egg tooth has scraped a hole in the shell.

DAY 2 2.20AM
The tortoise turns around inside its egg while it hatches, so it can break another part of the shell.

CRAWLING AROUND
The young tortoises crawl away from the nest and soon start to eat. From a young age, they eat the same grasses and leaves as their parents.

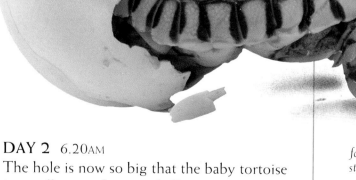

DAY 2 6.20AM
The hole is now so big that the baby tortoise just walks out of its eggshell.

The baby tortoise's shell is still soft and folded, but it will soon straighten and harden.

TORTOISE eggs can be hard or soft. The leopard tortoise lays hard eggs. When the female is ready to lay her eggs, she digs a hole in the ground with her back feet, drops the eggs in the hole, and covers them with soil. There are 10 to 25 eggs in each clutch. The eggs hatch more quickly in warm ground. At 86°F, they will hatch in about five months.

The baby tortoise takes a good look before it goes any farther.

DAY 2 4.30AM
The young tortoise bites off pieces of the shell to make the hole larger.

DAY 2 6.15AM
By pushing hard with its front legs, the baby tortoise manages to break out of the shell.

These two tortoises are newly hatched.

The spots on the leopard tortoises' shells are clearly visible. This is how they get their name.

CORN SNAKE

THE CORN SNAKE lives in North America. This colorful reptile is often kept as a pet because it is harmless to people. The name comes from its habit of living in cornfields. Corn snakes often climb trees and eat birds' eggs and nestlings. The female lays 8 to 16 soft eggs, often in a rotting tree stump.

CORN SNAKE eggs are 1¹⁄₃ in. long on average

Egg white is pouring out.

DAY 1 8.30AM
After 6 to 8 weeks incubation, the snake slits the shell.

DAY 1 9.15AM
The baby snake spends the whole day peeping out.

The egg white has now dried up.

DAY 1 11.30AM
The baby snake puts its head out and then goes back in again.

DAY 2 9.00AM
Suddenly the baby snake slithers out and glides away.

One day after first breaking the shell, the baby snake comes out of a second hole in the shell.

The young snake checks its surroundings with its tongue.

SLITHERING OFF
The young corn snake never sees its parents. It lives on the yolk in its body for a few days before starting to hunt frogs and lizards.

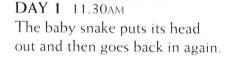

The snake is newly hatched. It will shed its skin when it is a few days old.

LEOPARD GECKO

THE LEOPARD GECKO is a kind of lizard. It lives in the grasslands of India, Pakistan, and Iran, and makes its nest in a hole in the ground. Geckos come out to hunt at night, and some even come indoors. The leopard gecko hunts insects. It stalks its prey and then pounces, just like a cat hunting a mouse.

Most LEOPARD GECKO eggs are about 1 in. in length

Split in shell

Sand sticking to the shell

The baby gecko takes its first look at the outside world.

DAY 1 1.25PM
Geckos lay two soft white eggs. This shell has just split open.

DAY 1 1.30PM
The gecko is starting to push its head out of the opened egg.

DAY 1 1.55PM
The baby gecko's head pops out of the hole in the egg.

In just 40 minutes, the baby gecko is free of its shell.

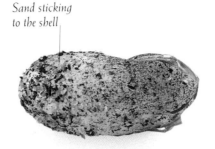

DAY 1 2.03PM
The baby gecko rests before its final efforts to struggle free.

DAY 1 2.04PM
After two to three months inside, the gecko walks out of the egg.

The young gecko is now ready to start its new life.

The gecko has just hatched out. The stripes on its body gradually change into spots as it gets older. It gets its name because of its spotted skin.

YOUNG AND FREE
The young gecko is independent as soon as it has hatched out.

BUTTERFLY, LADYBUG, & DRAGONFLY

INSECTS lay many tiny eggs. Most insect eggs, such as ladybug or butterfly eggs, hatch first into larvae, called caterpillars or grubs. The larvae look very different from the adults. Other insects, such as dragonflies, hatch into nymphs, which look like very small, wingless adults.

BUTTERFLY egg: 1.5mm long
LADYBUG egg: 2mm long
DRAGONFLY egg: 4mm long

The eggs are normally laid one at a time, on a plant that the larvae eat.

DAY 1
The swallowtail butterfly lays its bright yellow egg.

The larva, called a caterpillar, is starting to emerge.

DAY 10 5.00AM
The caterpillar uses its jaws to cut a hole in the eggshell.

DAY 10 7.00AM
As the caterpillar comes out of the egg, its body swells.

The caterpillar's bright colors warn birds to keep away.

DAY 18
The caterpillar feeds on leaves and grows bigger.

The eggs are laid in batches of about 15 to 20 eggs.

DAY 1
The ladybug lays a group of eggs, often under a leaf.

The eggs turn brown before hatching.

DAY 6
The eggs change color just before they hatch.

DAY 7
The eggs hatch together and the larvae crawl away.

The larva does not have wings.

DAY 21
The larva eats a lot of food and grows bigger.

The eggs are often pushed into stems or rotten wood for protection.

YEAR 1
DAY 1
The dragonfly lays its egg. The egg is usually well hidden.

YEAR 1
DAY 180
The nymph hatches out and breathes with gills.

The stripes deter larger nymphs from eating this one.

YEAR 1
DAY 190
The nymph hunts and eats other small pond animals.

The nymph may tackle prey as large as a tadpole.

YEAR 2 DAY 1
The nymph has grown much bigger now, but it still has no wings.

NYMPHS develop wings and grow into adults. Larvae go through another stage first. They rest in a safe place and turn into pupae. Inside the pupa, or chrysalis, the insect's body changes into the adult shape. Eventually, the skin splits and the adult insect emerges, as if hatching from an egg.

The chrysalis is fixed to the twig by a silky thread.

DAY 35
The caterpillar slowly turns into a chrysalis.

Inside the chrysalis, the caterpillar has changed into the adult butterfly.

DAY 50
The fully formed adult butterfly has just emerged from the chrysalis.

SWALLOWTAIL BUTTERFLY

The butterfly's wings are now open, dry, and brightly colored.

The pupa is normally attached to a leaf.

DAY 28
The larva gradually changes into a pupa.

Discarded pupa case.

DAY 35
The pupa case splits and the adult ladybug emerges.

The ladybug is yellow at first.

DAY 36
It takes 24 hours for the ladybug to turn red.

7-SPOT LADYBUG
The adult ladybug will lay her own eggs within a few weeks.

The nymph has crawled out of the water.

YEAR 2
DAY 320
The nymph's skin has split and the adult dragonfly is coming out.

The nymphal skin is left behind.

One hour later, the adult is out.

It takes one week for the adult to become brightly colored.

SOUTHERN HAWKER DRAGONFLY
Dragonflies only live for up to eight weeks. During this time, they must breed and lay their eggs to continue the cycle.

COMMON FROG

FROGS ARE AMPHIBIANS. These are animals that live first in water, then on land. The adults spend most of their time on land, but every spring they find a pond and lay their spawn, or eggs, in the water. The eggs hatch into larvae, called tadpoles, which live in the water until they grow into froglets.

FROG eggs are 3mm in length on average

The eggs have just been laid.

This tadpole is just hatching out.

Tadpoles look very different from adult frogs.

DAY 1 11.30PM
Each frog egg is surrounded by a ball of transparent jelly.

DAY 10 10.15AM
The tadpole has formed and is almost ready to hatch out.

DAY 17 2.30PM
Newly hatched tadpoles are nourished by their yolk remains.

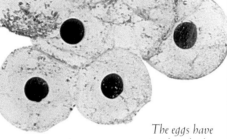

The tadpoles' tails will shrink back as their legs grow.

Front legs

Back legs

DAY 43
After six to eight weeks, the tadpoles' back legs grow.

DAY 53
The front legs have now grown as well, and the tadpoles are beginning to look like frogs.

FULLY GROWN
It takes the female frog three years to grow up and lay her first eggs. Each frog lays about 1000 eggs at a time. Very few of these eggs survive and grow into adults because tadpoles and frogs have so many predators.

GREAT CRESTED NEWT

NEWTS ARE AMPHIBIANS that belong to the salamander group. The adults always look like tadpoles because they keep their tails and their legs are very small. They live in damp places on land for much of their lives. Unlike frogs, newts stay in the water for several weeks after they have laid their eggs.

> The average length of a NEWT egg is 3mm

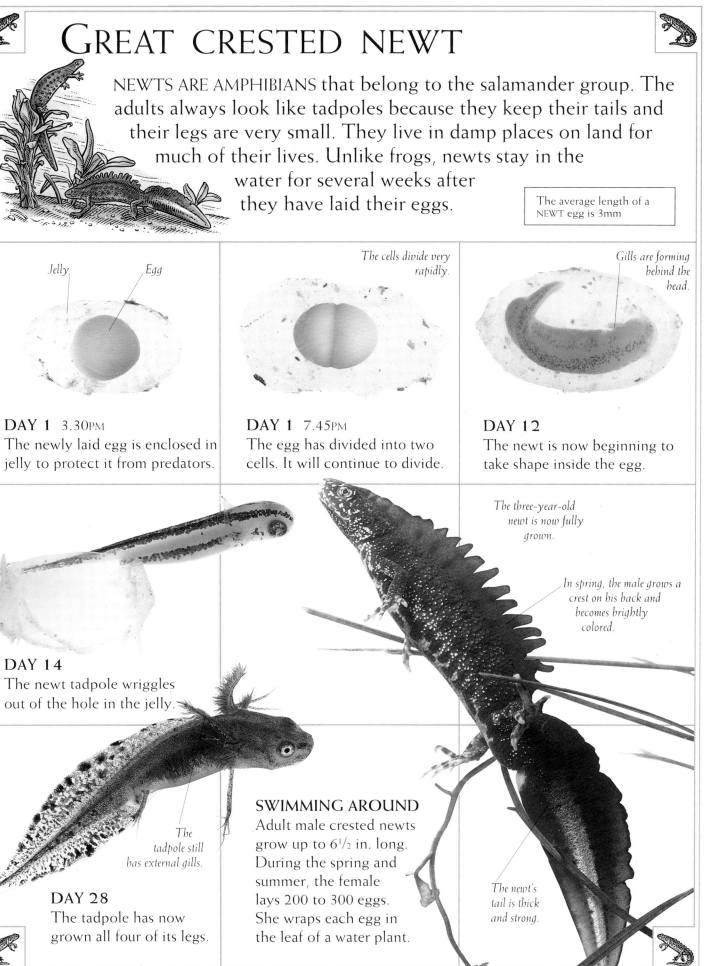

Jelly *Egg*

DAY 1 3.30PM
The newly laid egg is enclosed in jelly to protect it from predators.

The cells divide very rapidly.

DAY 1 7.45PM
The egg has divided into two cells. It will continue to divide.

Gills are forming behind the head.

DAY 12
The newt is now beginning to take shape inside the egg.

DAY 14
The newt tadpole wriggles out of the hole in the jelly.

The three-year-old newt is now fully grown.

In spring, the male grows a crest on his back and becomes brightly colored.

The tadpole still has external gills.

DAY 28
The tadpole has now grown all four of its legs.

SWIMMING AROUND
Adult male crested newts grow up to 6½ in. long. During the spring and summer, the female lays 200 to 300 eggs. She wraps each egg in the leaf of a water plant.

The newt's tail is thick and strong.

41

DOGFISH, TROUT, & GOLDFISH

THERE ARE more than 20,000 varieties of fish and they lay their eggs in many ways. Some fish, such as cod and flounder, lay millions of small eggs that float in the sea. Once hatched, the baby fish also float until they can swim properly. Other fish lay fewer, larger eggs.

DOGFISH egg: 2½ in. long
TROUT egg: ¼ in. long
GOLDFISH egg: 2mm long

The baby dogfish is visible through the horny case.

DAY 210
10.00AM
The fish has been developing inside its eggcase for seven months.

The fish pokes its nose out of the eggcase.

DAY 240
6.00AM
Eight months after the egg was laid, the dogfish is ready to hatch out.

The fish is coming out headfirst.

DAY 240
6.05AM
The baby dogfish wriggles its tail to get out of the tough eggcase.

The fish looks too big to have fitted in the case.

DAY 240   6.09AM
After just nine minutes, the fish is almost out.

Blood vessels have already developed.

DAY 14
Two weeks after the egg is laid, the fish's body starts to take shape.

The eyes are clearly visible.

DAY 25
The baby trout is almost ready to wriggle out of the egg.

The fish is emerging. It will hatch sooner in warmer water.

DAY 28 9.45AM
The fish has broken its soft case and is coming out quickly.

The baby fish has plenty of yolk when it hatches out.

DAY 28 9.46AM
The baby trout, called an alevin, has just emerged.

The eggs normally stick to water plants.

DAY 2
Two days after the egg is laid, the fish has started to develop.

The fish's beady eye can be seen clearly.

DAY 6
The baby goldfish is now almost fully formed and will soon hatch out.

Hatching time depends on the water temperature. It varies from 5 to 10 days.

DAY 7
After seven days, this goldfish is now hatching out of its egg.

DAY 14
The young fish starts to eat when its yolk is used up.

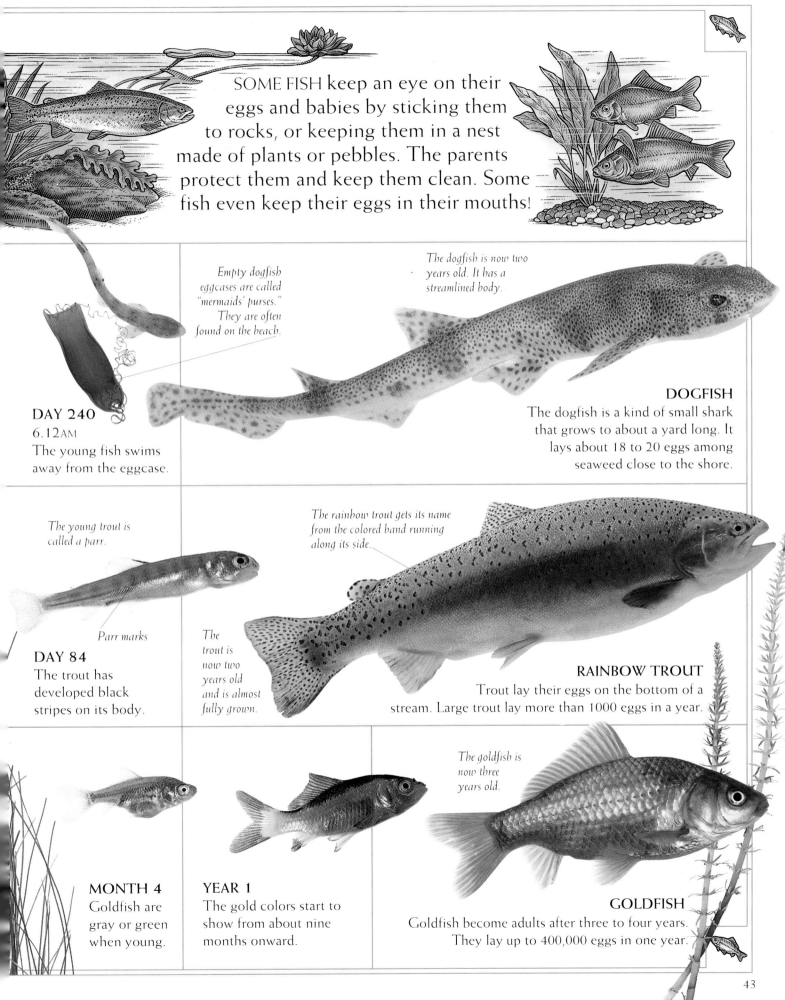

SOME FISH keep an eye on their eggs and babies by sticking them to rocks, or keeping them in a nest made of plants or pebbles. The parents protect them and keep them clean. Some fish even keep their eggs in their mouths!

Empty dogfish eggcases are called "mermaids' purses." They are often found on the beach.

The dogfish is now two years old. It has a streamlined body.

DAY 240
6.12AM
The young fish swims away from the eggcase.

DOGFISH
The dogfish is a kind of small shark that grows to about a yard long. It lays about 18 to 20 eggs among seaweed close to the shore.

The young trout is called a parr.

The rainbow trout gets its name from the colored band running along its side.

Parr marks

DAY 84
The trout has developed black stripes on its body.

The trout is now two years old and is almost fully grown.

RAINBOW TROUT
Trout lay their eggs on the bottom of a stream. Large trout lay more than 1000 eggs in a year.

The goldfish is now three years old.

MONTH 4
Goldfish are gray or green when young.

YEAR 1
The gold colors start to show from about nine months onward.

GOLDFISH
Goldfish become adults after three to four years. They lay up to 400,000 eggs in one year.

KERRY SLUG

SLUGS LIVE in damp places and come out mainly at night. The kerry slug lays around 80 eggs in total, in small batches over two to three months. The eggs are laid in holes in the ground or under rotting logs. They must stay moist – otherwise they will die.

SLUG eggs are 5mm in length on average

The baby slug is visible inside the egg.

DAY 1 11.20AM
A slug egg starts as a shiny oval ball. This egg is ready to hatch.

Split in eggshell

DAY 1 11.30AM
The egg hatches six weeks after it was laid. The slug starts to push itself out.

The slug has eyes on stalks on top of its head.

DAY 1 11.31AM
The baby slug emerges headfirst. It takes its first look at its surroundings.

The eggshell is completely transparent.

The slug has a slimy foot running along the underside of its body.

DAY 1
11.32AM
It is almost out of the shell and is already crawling away.

DAY 1 11.33AM
The slug is now out of its shell, which it leaves behind as it slithers off.

Slugs eat plants which they scrape with their rough tongues.

SLIDING AROUND

The slug is one year old and is now an adult. Slugs are like snails without shells. In fact, they do have tiny shells hidden under their skin. Their soft bodies are covered in a layer of slime which helps protect them. They leave behind a shiny trail of slime when they crawl over the ground.

INDEX

GLOSSARY

Albumen The "white" around the yolk of an egg. It helps protect the embryo.
Amnion The water-filled membrane that encloses an embryo.
Brood A group of young animals that hatch at the same time from the same clutch of eggs.
Camouflage Eggs and animals that are hard to see because their dull colors and patterns blend with their surroundings.
Cell The unit that makes up all living things.
Clutch A group of eggs that a bird lays at the same time.
Crèche A group of young animals from several families, that are guarded by only a few of their parents.
Down A baby bird's first fluffy coat. Most of the down is replaced by thicker feathers when the bird grows up.
Eggshell The outer layer of an egg, which can be hard or soft.

Egg lining The thin, soft layer inside the eggshell.
Egg tooth A hard lump on the tip of the beaks of baby birds and the noses of some other animals, which helps them break the eggshell from the inside. It falls off soon after they have hatched.
Embryo A baby animal while it is developing inside the egg.
Fertilization The joining of the male sperm with the female egg, so that the egg can start to develop into a baby animal.
Hatch Eggs do this when the animal starts to break out from inside.
Hatch out The baby animals do this when they break out of their eggs.
Incubation The period of time from when an egg starts to develop to when it hatches. Birds' eggs are kept warm by their parents during this period.

Larva The young stage of insects and some other animals, when they look quite different from their parents.
Membrane A thin sheet of living tissue.
Nestling A baby bird that is being looked after in the nest.
Nymph The young stage of an insect, such as a dragonfly, when it looks like a small, wingless version of its parents.
Ovary The organ in the female's body where eggs are made.
Pip The first break in the shell made from the inside of the egg by the baby animal.
Pupa The stage in the life of an insect during which the larva turns into an adult.
Yolk The part inside the egg that contains food for nourishing the embryo.

ACKNOWLEDGMENTS

Dorling Kindersley would like to thank the following people and organizations for supplying some of the animals photographed in this book: Ivan Lang of Arundel Wildfowl & Wetlands Trust, Hennie Fenwick of the British Chelonia Group, Dr Mike Majerus, Kay Medlock, Elizabeth Platt, Brighton Sealife Centre, Duncton Mill Hatchery, Tisbury Fish Farms, and the Reptile-arium, Enfield.

Additional photography by:
Neil Fletcher, Geoff Brightling, Frank Greenaway, Jerry Young, Cyril Laubscher and Harry Taylor.

Additional illustrations by:
Sandra Pond and Will Giles

Jane Burton and Kim Taylor would like to thank:
Rob Harvey of Birdworld, Farnham, Surrey for the ostrich, Humboldt penguin and crowned plover eggs; Mrs Fleur Douetil and Roy Scholey of Busbridge Lakes, Godalming,

Surrey for the Roman goose and black swan eggs; Michael Woods for the golden pheasant and Aylesbury duck eggs; Ashmere Fisheries, Shepperton for the quail egg and Robert Goodden of Worldwide Butterflies, Sherbourne, Dorset for the butterfly egg. All other species are from the Burton-Taylor aviaries and garden. The incubators used were from A.B. Incubators Ltd., Stowmarket, Suffolk.